Aliens Among Us: The Evidence

Visitors from Outer Space

Cherry Lake Press

Published in the United States of America by Cherry Lake Publishing Group
Ann Arbor, Michigan
www.cherrylakepublishing.com

Reading Adviser: Beth Walker Gambro, MS, Ed., Reading Consultant, Yorkville, IL

Photo Credits: page 1: ©Rastan / Getty Images; page 5: ©Zachary Byer / Shutterstock; page 6: ©Feoffer / Wikimedia; page 9: ©Steven Puetzer / Getty Images; page 10: ©Ray Massey / Getty Images; page 11: ©handout / Getty Images; page 13: ©vchal / Shutterstock; page 15: ©European Southern Observatory / M. Kornmesser / nasa.gov; page 16: ©RANDY ELI GROTHE/KRT / Newscom; page 19: ©Image Hans Elbers / Getty Images; page 20: ©by Marc Guitard / Getty Images; page 21: ©Matt84 / Getty Images; page 22: ©Kevin Schafer / Getty Images; page 25: ©JPL-Caltech / nasa.gov; page 26: ©3Dstock / Shutterstock; page 29: ©Staff / Getty Images; page 30: ©Fotovika/Shutterstock

Cherry Lake Press is an imprint of Cherry Lake Publishing Group.

Library of Congress Cataloging-in-Publication Data has been filed and is available at catalog.loc.gov.

Cherry Lake Publishing Group would like to acknowledge the work of the Partnership for 21st Century Learning, a Network of Battelle for Kids. Please visit *http://www.battelleforkids.org/networks/p21* for more information.

Printed in the United States of America

Mari Bolte has worked in publishing as a writer and editor for more than 15 years. She has written dozens of books about things such as science and craft projects, historical figures and events, and pop culture. She lives in Minnesota and loves looking up at the stars.

CONTENTS

CHAPTER 1

Aliens All Around

Off the coast of Malibu, California, a strange stadium-shaped structure is hidden 2,000 feet (610 meters) underwater. It can be seen using Google Earth. It looks like a flat-topped oval and is nearly 3 miles (4.8 kilometers) wide. People call it Sycamore Knoll. They also say unidentified flying objects (UFOs) have been seen coming and going from Sycamore Knoll for 40 years. One UFO researcher claims that between 50 and 100 people have seen UFOs around the area. They believe it's a secret alien base. Others say it could be a **portal** to outer space. Maybe **extraterrestrials** have found a way to come and go without attracting attention in the sky.

Alien and UFO sightings aren't new. In 1897, large mysterious ships were seen over California, Texas, and the Midwest. In 1952, "flying saucers" were seen over

Fencing, razor wire, and barbed wire are meant to keep curious people away from Area 51.

EYES ON AREA 51

Sycamore Knoll's Google Earth **coordinates** are 34° 1'23.31"N 118° 59'45.64"W. Another alien-related location can be seen via Google Earth too. Area 51 is in the deserts of Nevada. It has long been believed to hold secrets about extraterrestrials. The coordinates for Area 51 are 36° 38' 24"N, 116° 24' 0"W. Satellite views are the only way you can get close to Area 51. The closest public road to the top-secret military base is 26 miles (42 km) away. The airspace is closed over Area 51 in a 575-square-mile (1,489-square-km) rectangle overhead.

Gray rubber strips, tinfoil, tape, heavy-duty paper, and sticks were part of the Roswell debris.

Washington, D.C. A few years earlier, in 1947, **debris** from a mysterious flying object was discovered in Roswell, New Mexico. In 1969, future president Jimmy Carter reported seeing a UFO in Leary, Georgia. In 2004, U.S. Navy pilots saw strange objects flying through the sky. They were reported again 10 years later.

For a long time, scientists have been waiting for a sign that aliens are out there. The Search for Extraterrestrial Intelligence (SETI) has been officially recognized since 1960. Some scientists are even sending messages into space and hoping for an answer back. The National Aeronautics and Space Administration (NASA) is scanning the sky for proof of Earthlike planets that could host life. They are also preparing astronauts here on Earth to spend more time among the stars.

Some people believe aliens have been around us for a long time. There are many mysteries here on Earth that are still unexplained. Disappearances, strange objects on the ground and in the sky, and even close encounters point to proving extraterrestrials have already visited us. If they've been here, why haven't they made direct contact? Or *have* they? Perhaps aliens have tried to let us know they're here, and we just haven't gotten the message yet. Maybe we just need to look more closely at the evidence we already have.

TAKEN

The first widespread alien abduction story in the United States happened to Betty and Barney Hill in 1961. They were driving home at 10 p.m. when they saw a bright light in the sky that began to follow them closely overhead. Barney saw humanlike figures wearing black. They tried to drive away. Then they heard buzzing and beeping noises. Their bodies began to tingle. Then they passed out. When they woke up, they were 35 miles (56.3 km) from their original location. They had no idea how they got there. Both say they recall being taken onto a spaceship and being examined. Their story has never been proven. But even if it remains a mystery, other people have been inspired to believe in aliens.

CHAPTER 2

Saying Hello

America's first recorded UFO sighting dates back to 1639. John Winthrop, the governor of the Massachusetts Bay Colony, wrote about it. He said that three men in a boat at night had reported seeing a light in the sky. The light moved back and forth over a distance of around 2 miles (3.2 km). Other people saw it too. When the lights faded, the men found themselves 1 mile (1.6 km) upstream. They had no idea how they had gotten there.

In the late 1940s, the U.S. government started becoming concerned about all the UFO reports they were receiving. **Astronomer** J. Allen Hynek was asked to review the sightings. By the 1960s, he was known as the world's expert on UFOs.

The Grays, also known as Zeta Reticulans and Roswell Grays, are reported to be small, humanlike gray aliens with large heads and big black eyes.

After decades of talking to people about their UFO experiences, Hynek began to notice a pattern. The people who came forward were reluctant to do so. They were respected members of their communities. They had no reason to make up a story. Hynek believed their experiences were real. In 1972, he wrote a book called *The UFO Experience*. In it, he classified the different types of close encounters with extraterrestrials.

THE RIO SCALE

The Rio Scale was introduced during the 51st International Astronautical Congress in 2000. It was officially adopted in 2002. It assigns alien encounters a score on a scale from 0 to 10. A score of 0 means the encounter was unimportant or insignificant. Things like human-made signals, natural phenomena, or a **hoax** would be a 0. A score of 10 means it was extraordinary. Things like reliability, how it was observed, and the quality of reporting help define the score. So far, no events have scored greater than a 1. But the scale will make it easy to classify future events and real encounters.

J. Allen Hynek was involved with the air force studies Project Blue Book, Project Grudge, and Project Sign. All three were scientific studies involving UFOs.

- Close encounters of the first kind happen when someone sees a UFO. The UFO does not leave any evidence it was ever there.
- Close encounters of the second kind happen when a UFO leaves physical evidence it was there. Broken tree branches or burn marks on the ground are examples.
- Close encounters of the third kind happen when a person sees an alien or makes contact with the UFO or its occupants.

Since Hynek's theories, others have defined two more kinds of close encounters. Close encounters of the fourth kind are actual alien **abductions**. Close encounters of the fifth kind include communication between humans and aliens.

Hynek worried that **skeptics** would prevent people from coming forward to talk about their own experiences. Many reports of UFOs have been explained by science. They usually turn out to be weather events or human-made aircraft. But between 10 and 20 percent of the reports remain unexplained. That number could be even higher if more sightings were reported.

THE NATIONAL UFO REPORTING CENTER

The National UFO Reporting Center was founded in 1974 by a ufologist. The center records and looks into reports from people who have seen UFOs. More than 150,000 reports have been taken. Callers can remain anonymous, which removes some of the **stigma** from reporting. All reports can be viewed by the public for free. They are sorted by state, event date, date posted, and the shape of the UFO.

TYPES OF ALIEN VISITORS

Visitors have long been reported on Earth. Believers and **cryptozoologists** have gathered accounts of sightings over the years. Here are a few of the more common aliens that have been reported.

GRAYS

The Grays are short, gray-skinned beings with large heads and big, black eyes. Grays are the most commonly reported sightings since the 1960s.

NORDICS

Tall with blond hair and blue eyes, Nordics could be from another dimension. Others believe the Nordics are here to help humans.

REPTILIANS

Lizard-like creatures can take on any form thanks to shapeshifting and holograms. They are exploring Earth as a potential outpost to their planet.

CRYPTIDS

These are legendary animals; Bigfoot, the Loch Ness Monster, Chupacabra, and the Jersey Devil are all believed to be **cryptids**.

CHAPTER 3

Tracking the Visitors

Humans create a lot of trash and pollution here on Earth. If aliens are advanced enough to travel through space, chances are they will create those things too. In 2017, a fast, cigar-shaped object flew past Earth. It was the first object seen in our solar system that came from somewhere else. Researchers called it 'Oumuamua (oh-MOO-uh-MOO-uh).

More than 27,000 pieces of space junk **orbit** our planet. The junk is from spacecraft that are no longer working, old launch equipment, or broken satellites. In 2017, astronomer Avi Loeb suggested that 'Oumuamua was space junk from an alien world. Researchers theorized that it was potentially a part of a light sail—a spacecraft

'Oumuamua is the first known interstellar object to visit our solar system. *Interstellar* means the space between stars.

ALIEN DREAMS

Doctors and psychologists use science to study reported sightings and abductions. Harvard doctors used hypnosis on people who said they had been abducted by aliens. Some doctors believe that hypnosis may help people remember things more clearly. The doctors found that remembering the supposed abduction caused a strong physical reaction, similar to what other people who have had traumatic life experiences feel. To explain this reaction, the doctors suggest it may have been a result of sleep paralysis, when people wake from dreams before they can move their bodies. They end up seeing flashing lights, hearing noises, and having hallucinations. The human brain tries to explain what a person thought they saw or felt. For some people, that explanation is extraterrestrial.

Walt Andrus is the founder of MUFON.

that uses sunlight to move. It could also be a **probe**. Unfortunately, 'Oumuamua disappeared before scientists could do much study on it.

The Mutual UFO Network (MUFON) is the oldest and largest UFO organization in the world. Its goal is to investigate UFO sightings and collect them into a database. Many governments have looked into UFOs and decided they are not a threat to us. They do not see the value in spending money to do further research. MUFON fills that gap. It holds conferences where people can present new findings. It also has more than 600 MUFON UFO investigators to take reports.

The U.S. government is starting to publicly recognize UFOs, though. In 2021, a report was released. It summarized 143 reports of UFOs (also known as unidentified aerial phenomena, or UAP) received since 2004. Many of the reports were made by members of the U.S. Navy.

In 2022, the National Defense Authorization Act was signed. It calls for a new government program. The program will study and investigate UFO sightings. They will also keep Congress and other nations updated on what they find. Service members who suffer from health issues after encountering UFOs will receive any medical care they need.

Some people are skeptical that the act will lead to formerly secret information being made public. Others believe that this reinforces the idea that UFOs could be threats to humanity.

UFOLOGY

Ufology is the investigation of UFOs. It may sound like a made-up thing or a fun hobby. But some ufologists have completed higher education. One-fifth have higher degrees such as PhDs. Ufologists range from engineers and physicists to teachers and government workers. Some online universities offer degrees in ufology. Other people end up working for MUFON or the British UFO Research Organization. Both programs require training and certification.

CHAPTER 4

Signs of Life

Some believers think that aliens have already left behind physical proof of their presence. Stone formations and crop circles are examples of things that seem too heavy or too technologically advanced for ancient people to have created. Some things, such as the pyramids of ancient Egypt and the Nazca Lines that cover the sands of Peru, are believed to have been created thanks to alien help. Many people believe that these things don't prove aliens were here. But they still remain unexplained.

Stonehenge in England is probably the most famous circle of stones on Earth. About 100 huge stones stand together. Some of them weigh as much as 25 to 30 tons and came from about 25 miles (40 km) away. They are perfectly arranged to line up with the sunrise on the first day of summer.

About a million people visit Stonehenge each year.

Some of the straight Nazca Lines run as far as 30 miles (48 km) in one direction.

In 2013, the British Ministry of Defence released a huge stack of documents related to UFOs. More than 4,300 pages described strange sights over Stonehenge, Parliament, and other locations across the United Kingdom. Because they didn't appear to be a threat, the government never followed up on them.

People who fly over Peru will see strange shapes if they look at the land. White lines rise from a dry, red desert. They cross over each other and move in swirls and zig-zags. In total, there are more than 800 straight lines, 300 shapes, and 70 plant and animal shapes.

UNIDENTIFIED SWIMMING OBJECT

The Loch Ness Monster has been a mystery since a sighting made the news in 1933. Since then, many scientific investigations have been made to find the cryptid supposedly living in a lake. The most famous photo of "Nessie," taken in 1934, has been proven to be a fake. Yet, people still want to believe. Some believers suggest Loch Ness might be a portal to another time or place. In 2015, tourists noticed two UFOs flying over the lake. Some people may want to believe that other aliens might be using the portal too. It's important to document sightings, even if they are made up. Hearing about other peoples' experiences helps us explain our own unexplained incidents.

Peru does not have much wind or rain. This has protected the Nazca Lines for hundreds of years.

Visitors to Peru couldn't see the Nazca Lines until the 1930s, when commercial flights began flying people across the country. The first archaeologist to study the lines called them "the largest astronomy book in the world." The lines are about 2,000 years old. But nobody knows why they were made. Some believe they were made to guide aliens to Earth.

Odd patterns in farmers' fields date back to 1678. A farmer woke up to neat circles about 1.5 acres (0.6 hectares) cut into his straw field. The circles were so perfect that it seemed impossible that someone could have made them overnight. Crop circles continued to pop up around the globe. In the 1960s, some were seen in Australia. Rumors spread that they were caused by UFO landings. A sugar cane farmer said he saw a UFO rise off the ground. When he went to check it out, he found the sugar cane reeds had been woven into perfectly circular shapes.

ENGINEERS ACROSS THE AGES

Questioning engineering marvels should be done with care. Many ancient civilizations were able to accomplish great things. Some historians feel that giving aliens credit for the unexplained diminishes the contributions of early civilizations. Stonehenge is an ancient structure on European soil. Most of the engineering marvels are in Africa, South America, and North America. Recognizing these peoples' contributions, while still wondering how unexplained structures were built, can be done at the same time.

CHAPTER 5

We're the Aliens

Letting aliens know that we're here and that they can visit at any time has been part of our space exploration. Launched in 1972 and 1973, the *Pioneer 10* and *Pioneer 11* space probes carried small metal plaques. Each marked the probes' place of origin and launch date. They also had a simple picture of a man and a woman. Any higher beings who find the probes can find their way here. *Pioneer 10* will reach its destination in 2 million years. *Pioneer 11* will pass near a star in 4 million years.

NASA sent the *Perseverance* rover to Mars in 2020. A metal plate is attached to *Perseverance* too. It holds silicone chips in place. The chips are etched with microscopic print with nearly 11 million names. The names belong to space fans who wanted to be part of the mission. The plate itself has a map of Earth, the Sun, and Mars. The rays of the Sun contain a message in **Morse code**. The message says, "Explore As One."

Perseverance touched down on the surface of Mars on February 18, 2021.

Tardigrades have eight legs and a head. When they are dehydrated, they pull their head and legs in and curl into a ball to stay safe. When water is reintroduced, they uncurl.

Chasing down aliens that might have stopped by for a glance at Earth is another possibility of making contact. The sighting of 'Oumuamua inspired others to think about ways to get to the mysterious rock and other strange objects traveling through space. The Initiative for Interstellar Studies (i4is) wants to make spaceships a reality in our lifetime. They launched a program called Project Lyra.

WATER BEAR RESCUE

In 2019, Israel sent a spacecraft to the Moon. It crashed. And there were living beings on board. Water bears, also called tardigrades, are tiny micro-animals. They live on every continent on Earth. They can survive temperatures as high as 302 degrees Fahrenheit (150 degrees Celsius) and as low as **absolute zero**. They were dehydrated for the trip. When the spacecraft crashed, they were scattered across the surface of the Moon. Most creatures need water to live. But water bears can live for decades after dehydration. There is not much water on the Moon, so it is unlikely they will wake up any time soon.

FURTHER DISCOVERIES

It might sound far-fetched to think that aliens are on Earth. But there are an estimated 8.7 million species on Earth—and we've only identified about 1.6 million of them. A million species are also under threat of extinction. More than 450 species have gone extinct in the past 10 years alone—and those are the ones we know about. Around 20 new species are named every year. In 2020, researchers officially discovered and named 213 new species. With so many continuing their life in secret, is it such a jump to wonder if aliens could be among them?

Project Lyra would find out if our current or near-future technologies could be used to meet up with 'Oumuamua. A light sail similar to what Avi Loeb thought 'Oumuamua itself was using seems to be the most likely way to do it. i4is hopes to be among the first to contribute to robotic and human exploration in space.

We could also help aliens hitch a ride back to Earth, whether we mean to or not. Humans have a long history of moving into new, unexplored places. When we send out

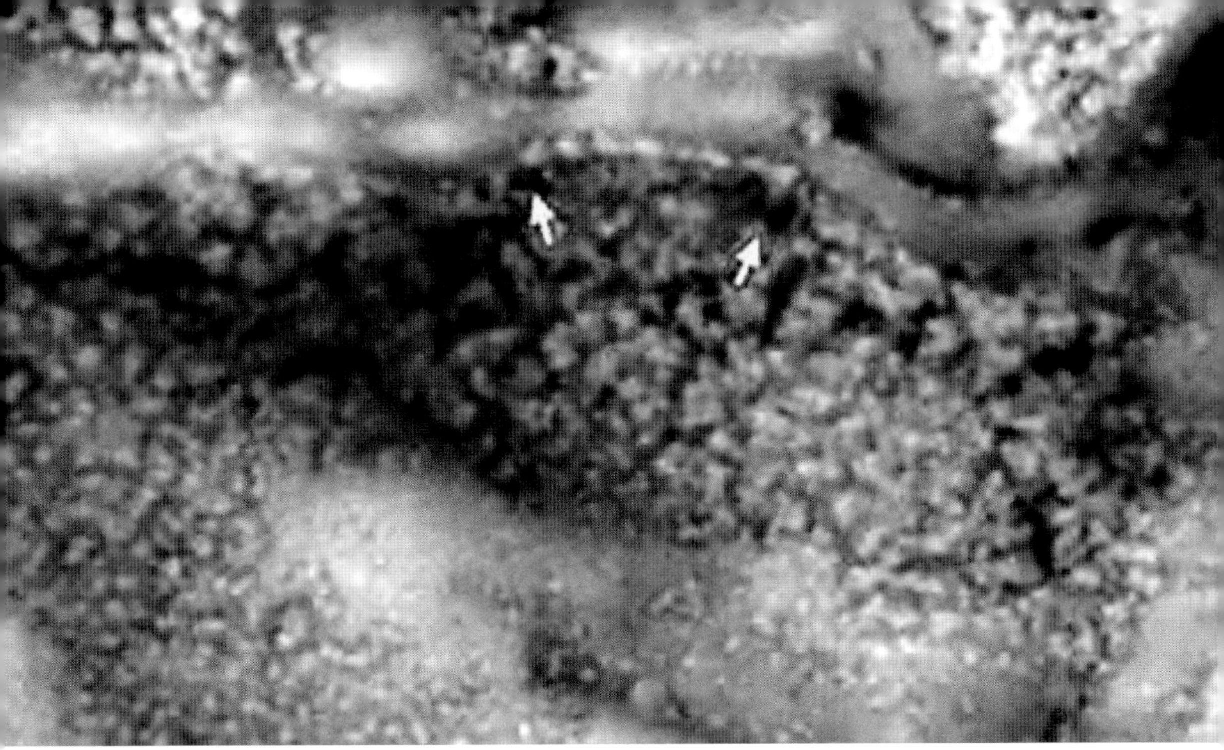

In 1996, a team of NASA scientists said they found fossilized microbes on a 16-million-year-old meteorite from Mars. Scientists today are still debating the claim.

spacecraft and probes, we could be sending our microbes along with them. And it's definitely a goal to send people into deep space someday. Aliens might have that same desire to move in. Right now, the known risk is tiny microbes. But that's because we don't have proof of anything bigger yet.

Whether aliens are watching us from space or are already here, humans are fascinated with the idea of them. Believers will continue to scan the sky for signs of life. Maybe one day, something will show up.

Activity

Tell Me About It

1. Choose your favorite place, activity, person, or animal.
2. Describe it as though you're describing it to an alien or someone who has never seen that thing or done that activity before. Be very specific.
3. Remember that things like comparisons or measurements may not mean anything to aliens because they have no frame of reference for those things. How hard was it to explain your object or place?
4. Now imagine you're trying to explain it to someone who also doesn't speak your language! What do you do differently to describe it this time?

Learn More

Books

Bolte, Mari. *Earth vs. Mars*. Ann Arbor, MI: Cherry Lake Publishing, 2022.

Hubbard, Ben. *What Do We Know About Crop Circles?* New York, NY: Penguin Workshop, 2022.

Morancy, Joalda. *Aliens: Join the Scientists Searching Space for Extraterrestrial Life*. New York, NY: Neon Squid, 2022.

Spilsbury, Louise. *Alien Visitations*. New York, NY: Crabtree Publishing, 2022.

On the Web

Curious Kids
https://www.space.com/would-alien-life-take-our-call-curious-kids

SETI Institute
https://www.seti.org

Tour of the Universe
https://tpt.pbslearningmedia.org/resource/buac18-912-sci-ess-tourofuniverse/tour-of-the-universe

Glossary

abductions (ab-DUKT-shuhns) the acts of being taken by force or deception

absolute zero (ab-suh-LOOT ZEE-roh) the lowest known possible temperature; absolute zero is minus 459.6 degrees Fahrenheit (minus 273.1 degrees Celsius)

astronomer (uh-STRAH-nuh-mur) someone who studies astronomy

coordinates (coh-OR-dih-nihts) a set of values that show an exact location

cryptids (KRIP-tids) creatures that are said to exist, but their existence has not been proven

cryptozoologists (KRIP-toh-zoh-AHL-uh-jists) people who study and search for legendary or extinct animals

debris (deh-BREE) scattered pieces of broken rock or waste

extraterrestrials (EK-struh-tuh-RES-tree-uhls) from outside Earth

hoax (HOHKS) a deception used to cover up the truth

Morse code (MORS KOHD) an alphabet or code where letters are represented by long and short signals

orbit (OR-bit) the curved path of an object around a star, planet, or moon

portal (POR-tuhl) a doorway, gate, or other entrance

probe (PROHB) an unpiloted spacecraft that travels through space to collect information

skeptics (SKEP-tiks) people who do not believe

stigma (STIG-muh) a negative and often unfair belief people have against something

Index